HOW TO STING THE
POLYGRAPH

DOUG WILLIAMS

Copyright © 2020 Doug Williams
All rights reserved
First Edition

All rights reserved. No part of this publication may be reproduced, distributed, or transmitted in any form or by any means, including photocopying, recording, or other electronic or mechanical methods without the prior written permission of the publisher.

WWW.POLYGRAPH.COM
PO BOX 720568
NORMAN, OK
73070
405-226-4856
doug@polygraph.com

CONTENTS

How To Sting The Polygraph 5

The "Pre-Test" Interview. 11

The (Polygraph) Test Phase 31

The "Post-Test" Interrogation 42

Practice Examinations 52

A Note to the Reader. 65

* Pre-Employment Pre-Test Questions 70

HOW TO STING THE POLYGRAPH

My name is Douglas Gene Williams. I was a detective sergeant with the Oklahoma City Police Department and ran the polygraph section of the Internal Affairs Unit for seven years. I administered thousands of tests for the Oklahoma City Police Department, and for many other agencies including the Secret Service and the FBI. I have also administered hundreds of tests as a private polygraphist.

I have the dubious distinction of being the only licensed polygraphist to ever tell the truth about the so called "lie detector". And the truth is, the polygraph is no more accurate than the toss of a coin in determining whether a person is telling the truth or lying – and I am proud to say, that is the exact same phrase the US Supreme Court used when they refused to allow the results of the polygraph into evidence.

While on the police department, I discovered that what I did for a living was a fraud, and in 1979 I resigned and embarked upon what turned out to be a crusade to outlaw the polygraph. I set about to prove three things about the polygraph that I knew to be true: (1) The polygraph test has a built-in bias against a truthful person, (2) It is certainly not capable of determining truth or deception, (3) Anyone can learn to control the test results so as to ALWAYS produce a "truthful" chart.

For example, regarding facts one and two, I offer this evidence. Some time back I was invited to prove these statements on CBS 60 Minutes. I spent about five weeks in New York City setting up a sting operation to prove that the polygraph has a built-in bias against a truthful person, and that it is certainly not capable of determining truth or deception. For this investigation, we set up a mock situation in a business setting. The setting was the offices of the magazine Popular Photography located in New York City. We cut holes in the walls of an office in the magazine's corporate headquarters and put our video cameras next door to secretly document the polygraph examinations, and we had our microphones in what appeared to be an overhead sprinkler system. We then picked three polygraph operators at random out of the yellow pages in New York City and hired them to test the employees of the magazine regarding the alleged theft of a camera. No camera had been stolen, but all three polygraph operators called honest, truthful people liars and thieves, and each one picked a different person! Talk about a sick joke, those guys showed what they were on national television with millions of people watching!

Unfortunately there continues to be literally millions of private stories about innocent people who have been branded as criminals by this machine. I rest my case on facts one and two.

Concerning fact number three, that anyone can learn to control the test results so as to ALWAYS produce a "truthful" chart, I have demonstrated this on every major network on national television over 20 times. For more information on this check out the media clips on my website **www.polygraph.com** . I have taught thousands of people how to control every tracing on the polygraph chart at will. It is ridiculously simple to master the skills necessary to pass the "lie detector" and produce what the polygraph operator should call a "truthful chart". I rest my case on fact number three.

I'm sure you already know the polygraph is a joke, but you don't want the joke to be on you. This manual is designed to teach you how to protect yourself by teaching you how the machine works and to show you how to control your reactions so that you will be able to produce a "truthful" chart.

Due to the fact that parts of this manual are somewhat technical, and you must read it at least four times in order to fully comprehend the information it contains, I have condensed it to as few pages a possible.

The word polygraph is derived from two Greek words: poly, which means many, and graphos, which means writings - many writings. The "many writings" which the polygraph records on its charts are you blood

pressure, pulse rate, respiration and galvanic skin response. Your blood pressure and pulse rate is recorded by the cardio pen which traces your heartbeat; this is referred to as the cardio tracing. Your breathing is recorded by the pneumo pens. This is referred to as the pneumo tracing. Your Galvanic Skin Response, which is basically the sweat or perspiration activity on your hand, is recorded and is referred to as the GSR tracing. (See Exhibit A)

Changes in your blood pressure, pulse rate, breathing, and sweat activity are referred to as reactions. These reactions appear on the polygraph chart as follows. (1) A pneumo reaction, which is simply a pen tracing up and down on the chart as you inhale and exhale. (2) A cardio reaction, which is the pen tracing your heart beat, and showing an increase or decrease in your blood pressure and pulse rate. (3) A GSR reaction which is nothing but a pen tracing of the increase or decrease of the sweat activity on your hand. In other words the polygraph operator can watch you breathe, watch your blood pressure and pulse rate go up and down, and watch your hand sweat. And on the basis of this, he presumes to be able to say whether or not you are a liar. How absurdly ridiculous!

The validity of the polygraph as a lie detector rests on the theory that physical changes, or reactions, are caused by the emotional stress of lying and cannot be manipulated. There is only one thing wrong with this theory, and that is it is just not true. In fact you can learn to control every tracing on the chart at will.

The problem with the polygraph is that the reaction the polygraph operators call a "lying reaction" can be and is caused by many things other than a lie. As a matter of fact any number of innocent stimuli can and do cause this exact same reaction. Fear, rage, embarrassment at having been asked a personal question, pain from the cardio cuff, even the tone of the examiner's voice can all cause the exact same reaction that the polygraph examiner would brand as a lie.

The polygraph is not a lie detector, and it is not a truth verifier, it is simply a crude reaction recorder, and the so called reactions it records can be caused by many things other than deception. I can even teach you how to duplicate this reaction by a simple breathing and muscle exercise. In fact, when you finish reading this manual, you will be able to control every tracing on the polygraph chart at will.

What is involved in a polygraph test?

Polygraph tests are divided into three phases: (1) The "pre-test" interview and "stim test", (2) The "test" phase, and, (3) The "post-test" interrogation. I will explain each of these in detail.

THE "PRE-TEST" INTERVIEW

In this phase, the polygrapher will tell you he just wants to get to know you, to get you to get everything off your mind. He may even tell you that all this is "just between you and me". Don't be fooled the polygrapher will report all admissions you make.

The pre-employment polygraph examination is the setting for some of the worst cases of polygraph abuse because, unlike the criminal suspect, the job applicant cannot refuse the test without suffering as a result of the refusal. It is unfortunate and ironic that applicants for government agencies requiring security clearances, and law enforcement applicants are about the only ones who still have to take a pre-employment polygraph. These polygraph examinations are simply interrogations, the only part the polygraph plays is to scare you sufficiently to insure the full disclosure of all the sordid details of past indiscretions. Do not tell the polygrapher anything that isn't already a matter or record and certainly do not admit to anything that could disqualify you. During the pre-test interview for pre-employment polygraph tests, the polygraph operator will ask you a series of questions, some of which are listed below. The whole point is to get you to make damaging admissions – so be careful how you answer.

* (These pre-employment pre-test questions are listed in the back of the manual)

The polygrapher may tell you the reason he is asking so many questions is to get to know you better so he can administer a good test. He may also tell you that passing the test is more important than any admissions you make, and that it will be to your advantage to tell the complete truth in order to pass the exam. He will exhort you to "get everything off your mind, discuss anything that is worrying you, so that nothing interferes with your polygraph test". Do not be deceived, these are merely interrogation tricks designed to try to get you to change your goal. What is your goal in taking a pre-employment polygraph examination? Your original goal is to get a job; you are trying to sell yourself to a prospective employer so you naturally put emphasis on your most positive attributes, your ability, training, education, energy, attitude, etc. The polygraph con man will try to change your goal by telling you that in order to get the job your must pass his test, and in order to pass, you must tell the complete truth to all the questions he asks. Your goal then will be to spill your guts in front of the hidden video camera while he sits back and prods you into disclosing information you would never tell another person. Do not allow the interrogator to turn the polygraph room into a confessional! Perhaps you cannot refuse to answer these questions, but you can control the amount of information you give. Do not buy into the polygraph operator's lie that he will respect the confidentiality of what you tell him, every word you say is being recorded and will be played back to the person paying for the test. Audio and video recorders and cameras, as well as the ever-present see-through mirrors on the wall, with the adjacent observation room to witness the proceedings, is all a part of the well-equipped polygraph suite.

The polygrapher will then explain "how the polygraph works" - it is all B.S. and they all memorize some version of this little speech. This is the textbook explanation that Department of Defense Polygraph Institute-trained polygraphers provide.

" You may be a little nervous, especially if you have not had a PDD ["psychophysiological detection of deception," a more scientific-sounding term for "lie detection"] examination before. This is expected and is quite normal. To help put you at ease, I will explain what the instrument is and how it works. The polygraph is a diagnostic tool that is used to determine if a person is telling the truth. It simply records physiological changes that take place in your body when you are asked questions. Today, changes in your respiration, sweat gland activity, and blood pressure will be recorded. Please notice the two rubber tubes on the desk. One will be placed across your chest and the other will be placed around your abdominal area. They will be used to record your breathing. There are two metal finger plates next to the rubber tubes. These plates will be attached to two of your fingers and will record your sweat gland activity. Finally, there is a blood pressure cuff on the desk. It is the same type of cuff a doctor uses to measure blood pressure. It will be placed on your arm and will monitor changes in your cardiovascular activity. These physiological changes are a result of an automatic response system in your body. It is a response system over which you have no control. For example, visualize yourself walking down a dark alley late at night. Suddenly you hear a loud noise. You will instantaneously decide either to remain where you are and investigate the source of the noise, or to flee the area, sensing danger to your well being. Regardless of the choice you make, your body automatically adjusts itself to meet the needs of the situation; your heart may beat faster, your breathing may change and you may break out in a cold sweat. When you were growing up, if you are like most people, you were raised to know the difference between right and wrong. Quite probably, all of the adults you came in contact with-- your parents, grandparents, relatives, teachers, church officials--taught you that lying, cheating, and stealing were wrong. Ever since you were a young child, you have been programmed to know that lying is wrong. Think about the first time you lied and got caught. Remember how your body felt during that confrontation. Your heart may have been racing or you may have been sweating.

However, the responses were automatic; your body adjusted to the stress of the situation. People are not always 100% honest. Sometimes it is kinder and more socially acceptable to lie than to be honest - such as telling someone you like their clothes when you really think the clothes are awful. It is important for you to understand that even though a lie might be socially acceptable or only a small lie, or a lie by omission, your body still responds. The recording on the polygraph will show only the physiological responses. It cannot know what kind of lie you are telling. Therefore, it is extremely important that you be totally honest..."

This whole speech is designed to scare and intimidate you. The polygrapher's explanation is deliberately false and he is trying to mislead you - telling a lie may or may not result in physiological changes measurable by the polygraph. When the polygrapher says, "It is important for you to understand that even though a lie might be socially acceptable or only a small lie, or a lie by omission, your body still responds," he really means, "It is important for me that you believe this to be true." Fear is an essential element of all polygraph "tests." In its 1994 assessment of the Ames case, the U.S. Senate Select Committee on Intelligence reports, "A former polygrapher noted that without proper preparation, a subject has no fear of detection and, without fear of detection, the subject will not necessarily demonstrate the proper physiological response." (U.S. Senate Select Committee on Intelligence, 1994) But the problem is that this fear of being falsely accused may also cause you to have a reaction that would result in truthful persons being accused of deception.

If you are taking a specific issue polygraph test. (criminal, fidelity, probationary, etc), the pre-test is more of an interrogation about the specific incident that you are accused of. Again, don't admit anything they don't already know. Keep insisting that you are telling the truth and that you look forward to taking the polygraph test in order to prove it.

Sometime during the pre-test interview phase, the polygrapher will conduct what is commonly referred to as a "stimulation test" or "stim test," - DoDPI calls it an "acquaintance test." He may tell you that the purpose is to allow him to "adjust the instrument" and to make certain that you are "capable" of physiologically responding if you were to intentionally tell a lie. This explanation is also B.S. – it is just another way to try to "psyche you out". The real purpose of the "stim test" is to fool you into believing that your polygrapher can read your mind and that your deception will be detected. The most common "stim test" is to let you pick a card out of a deck of supposedly randomly numbered cards. The polygraph operator will then instruct you to answer "no" to all the questions, even to the question about the card number you actually picked. He will then ask, "Did you pick card number fifteen?", "Did you pick card number seven?", and so on down the list of all the numbers on the cards. At the conclusion of the test, he will tell you, based on his analysis of the chart, which card you picked and therefore, which question you lied to. This is simply a trick. He has used two decks of cards. He makes a big deal of shuffling one deck in front of you, he will then divert your attention and change decks using one whose "random" numbers are in order. He has memorized the sequence of numbers, i.e. 15, 8, 3, 5, 7, 10, and 13, and he knows which card you have picked as soon as you pull it from the deck. Sometimes, the "stim test" is done with a deck of cards. The polygrapher will ask you to pick a card and not show it to him. Then, while you are connected to the polygraph, he will ask you to answer "no" to each question he asked. Suppose you draw the jack of diamonds. Your "stim test" might go like this: Did you pick a face card? (No.) Did you pick a number card? (No.) Your polygrapher nonchalantly tells you, "It's obvious you picked a face card." He then proceeds to ask: Did you pick a king? (No.) Did you pick a queen? (No.) Did you pick a jack? (No.) He then informs you, "The polygraph shows me you have drawn a jack." He continues: Did you pick a spade? (No.) Did you pick a club? (No.) Did you pick a diamond? (No.) Did you pick a heart? (No.) The polygrapher "analyzes" the charts and tells you, "It's clear you picked the jack of diamonds. No doubt about it. You're are an excellent candidate for the polygraph. You can't tell a lie without your body showing a reaction." But what your polygrapher wouldn't tell you is that you drew your card from a trick deck, in which every card is the jack of diamonds. Some polygraphers use a "known-solution" numbers "test," in which the polygrapher will ask you to pick a number, say, from one to nine, and to write the number you picked on a sheet of paper. The polygrapher will tell you to answer "no" each time as he asks, "Did you write 1? Did you write 2?," etc. And he will tell you that when you answer "no" to the number that you wrote - to deliberately lie and say "no." Did you write 1? (No.) Did you write 2? (No.) Did you write 3? (No.) Did you write 4? (No.) Did you write 5? (No.) Did you write 6? (No.) Did you write number 7? (No) Did you write 8? (No) Did you write 9 (No). Regardless of whether or not you showed any reaction when you lied about the number you picked, the polygrapher will attempt to convince you that you are not capable of lying without the polygraph instrument detecting it. This is how DoDPI instructed examiners to explain the "stim test" to volunteers in a recent research project:

"Administer a standard known solution numbers test-- using the rationale below. DO NOT show the test to the examinee, but convince the examinee that deception was indicated. NOTE: be sure to use the word acquaintance or demonstration test when discussing this with the examinee. I'm now going to demonstrate the physiological responses we have been discussing. This test is intended to give you the opportunity to become accustomed to the recording components and to give me the opportunity to adjust the instrument to you before proceeding to the actual test. In addition, this test will demonstrate to me that you are capable of responding and that your body reacts when you knowingly and willfully lie. The standard four components (two pneumograph tubes, electrodermal plates, and cardiovascular cuff) are attached at this time, followed by the acquaintance test. The acquaintance test should be conducted in the manner taught at DoDPI.... The results will be discussed with the examinee as follows: That was excellent. It is obvious that you know lying is wrong. You're not capable of lying without your body reacting. You reacted strongly when you lied about that number. Even though I asked you to lie and it was an insignificant lie, you still responded. That will make this examination very easy to complete as long as you follow my directions."

Play along with the polygrapher's silly game and do not try to subvert this test. When the polygrapher picks your card, you may turn it to your advantage by congratulating him on his expertise and telling him that you are now more confident than ever that the polygraph test will show you are telling the truth. NOTE: A secondary purpose of this "test" is to show the polygraph operator what type of a reaction you will have on the real test. Therefore, let us assume you picked card number five, or number five; you should manipulate a reaction when you answer "no" to this question. This will accomplish two things; first it will show him you are a "good subject", that you have the ability to react, and, second it will cause him to look for that specific type of reaction on the real test which is to follow.

The polygraph operator is confident of his technique, and you must be equally confident of yours. Don't be the polygrapher's punk. Don't let him "play" you. Learn how to "play" him. The polygraph cannot detect lies (it only records physiological data) and you can learn how to control every tracing on the chart.

Next, the polygrapher will tell you the questions that he is going to ask on the test. There are two types of questions - RELEVANT & CONTROL.

It is important that you recognize the difference between relevant and control questions in order to know when to manipulate or cause a reaction and when to control or stop a reaction. A relevant question is obviously one that pertains to the issue at hand, for example, if the polygraph test is about a specific theft, and then the relevant questions have to do with the specific item that was stolen and whether you stole it. The relevant questions are those that are relevant to why you are taking the test. If you are taking a pre-employment test, the relevant questions are those that decided whether you get the job.

The polygraph test is simply a comparison of your reactions. The polygraph operator will compare your reaction to the relevant question with your reaction to the control questions. If your reaction to the relevant question is greater than your reaction to the control question he will assume you are lying, if your reaction to the control question is larger, he will assume you are truthful. Obviously you should show the correct reaction to the control questions, and show no reaction whatsoever to the relevant questions. And you must PRACTICE until you can do it properly!

There are three types of control questions. We will start with the KNOWN-LIE CONTROL QUESTIONS since they are the most commonly used.

KNOWN-LIE control questions are those which the polygraph operator assumes you will respond to with a lie, for example, "Have you ever stolen anything?" The stress involved in your lying answer to this question will theoretically result in a reaction which is then compared to your reaction to the relevant question.

These known-lie control questions are matched to the situation, for example if you are a sex crimes suspect, or if you are taking a marital fidelity test, one of the known-lie control question may be, "Have you ever engaged in any unusual sex acts?" Known-lie control questions are different from the relevant questions because they are general in nature and nonspecific in terms of time. Another example of a commonly used known-lie control question is, "Have you ever stolen anything?". The polygraph operator will insist that you make some admissions to this question during the pre-test interview, and some will go to great lengths to "stimulate" you with this question. The polygraph operator will ask the basic question in the pretest interview, and you will make some minor admissions, admit a few minor childhood thefts, but do not say anything incriminating. The questions will then be reworded to, "Besides what you've told me, have you ever stolen anything else?" That will be the way the question is worded on the test. On the test itself, you will answer no and manipulate a reaction. Here are more examples of the known-lie control question:

Have you ever engaged in any unusual sex acts? You will make some minor admissions, then the question will be reworded, and on the test itself he will ask, "Besides what you've told me have you ever engaged in any unusual sex acts?" When this revised known-lie control question is asked on the test, you will answer "no" and manipulate a reaction.

Have you ever deliberately hurt another person? You will make some minor admissions, then the question will be reworded, and on the test itself he will ask, "Besides what you've told me have you ever deliberately hurt another person?" When this revised known-lie control question is asked on the test, you will answer "no" and manipulate a reaction.

Have you ever stolen anything? You will make some minor admissions, then the question will be reworded, and on the test itself he will ask, "Besides what you've told me have you ever stolen anything?" When this revised known-lie control question is asked on the test, you will answer "no" and manipulate a reaction.

Have you ever done anything that if discovered you would be ashamed of? You will make some minor admissions, then the question will be reworded, and on the test itself he will ask, "Besides what you've told me have you ever done anything that if discovered you would be ashamed of?" When this revised known-lie control question is asked on the test, you will answer "no" and manipulate a reaction.

These questions may start with the phrase, before the age of 21, before this incident, between the ages of 20 to 30, or something similar. The known-lie control

questions on your test may be worded somewhat differently but they will be similar to the examples listed above so you will be able to recognize them.

Here are some more examples of this type of known-lie control question.

Did you ever bring shame upon yourself or your family?

Are you the type of person who would lie to get out of trouble?

Did you ever cheat anyone out of anything?

Did you every lie to anyone in a position of authority?

Did you ever blame someone for something you did?

Did you ever cheat anyone out of anything?

Did you every lie to anyone in authority?

Did you ever tell a lie to someone who trusted you?

Did you ever lie to cover up something?

Have you ever told a lie that would get another person get into trouble?

Have you ever told a lie about a person, even if the person is telling the truth?

Have you ever lied to a loved one?

Have you ever taken something that does not belong to you?"

Since the age of 18, have you ever considered hitting someone in anger?

Have you ever lied to a supervisor?

Have you ever lied to loved ones?

Have you ever lied to parents, teachers, or the police?

Have you ever lied to get out of trouble?

Did you ever reveal anything told to you in confidence?

Did you ever cheat in school?

Did you ever cheat in college?

Did you ever betray the trust of a friend or relative?

Do you sometimes intentionally mislead or deceive your friends?

Are you a really honest person?

Are you absolutely trustworthy?

Do you think you are smarter than most people?

Are you an untrustworthy person?

Are you a dishonest person?

Have you ever done anything that would embarrass you if your parents found out?

Have you ever done anything you would be embarrassed to tell me about?

Have you ever lied about anything serious?

Memorize these questions so you can recognize them quickly. AND REMEMBER: IF THE POLYGRAPH OPERATOR IS USING KNOWN-LIE CONTROL QUESTIONS YOU ONLY REACT TO THE KNOWN-LIE CONTROLS. And you will know he is using the known-lie control questions because he will review them with you before the test.

When answering these questions, don't tell the polygraph operator anything he doesn't already know or can't find out on his own. Remember,, the polygraph test is the most important test any of you will ever take. Until you take one, you have no idea how traumatic and grueling it can be - it is that way for a reason. The polygraphers want you to be so frightened that you "spill your guts". Some federal agencies even give bonuses to the polygraph operators who get the most damaging admissions!

In fact, many people are so intimidated that they make statements that the polygrapher will use to disqualify or incriminate them - some people are so frightened that they confess to things they haven't even done! DON'T DO THAT! If you have anything to admit, put it on your application, or if you have anything to confess, tell it to the investigator. Please don't give the polygraph operator the satisfaction of bragging about what you told him that they could not have known without him and his magical machine! That is how these polygraph con men justify their existence - by getting you to make admissions to them that you haven't told anyone else!

The polygraph operator may ask you why you reacted to the control question, you need to make up some reason like, "I remember the look in my daddy's eyes when he found out I had stolen the harmonica.", or "I'm still embarrassed to talk about it.", or something such as that – use your imagination. Remember when he asks you about a control question rather than a relevant question you have already stung him. Most people flunk their polygraph tests because they say too much. You are being forced to take the test so it is up to you to be in charge of the amount of information you disclose. The polygraph operator is lying about the validity of the test, and you are under no obligation to take the test seriously. Most of the juicy information I gleaned during polygraph sessions was in response to illegal questions. Do not tell these perverse purveyors of purloined personal information anything more than is necessary to get through their 'trial by ordeal'! Do not believe anything they say about this being "off the record" or "just between us" – they will report every scrap of information they get from you!

In my humble opinion, everything about the "lie detector" is totally irrelevant, but the next category of control questions is even labeled as such. The irrelevant control questions are easily recognized because they are usually absurdly irrelevant. Some of the irrelevant control questions asked during the test may be:

Can you drive a car?

Do you smoke?

Is today Thursday?

Are you setting down?

Are the lights on in this room?

DO NOT REACT TO THE IRRELEVANT CONTROL QUESTIONS. IF THEY ARE THE ONLY CONTROL QUESTIONS BEING USED – DON'T REACT TO ANY OF THEM. IF THE POLYGRAPH OPERATOR IS USING BOTH

KNOWN-LIE CONTROLS AND IRRELEVANT CONTROLS
YOU ONLY REACT TO THE KNOWN-LIE CONTROLS.

The third category of control questions deal with the surprise stimulus or embarrassing personal questions. These are even more absurd than the irrelevant question, and may include some of the following: "What is the tenth letter in the alphabet?", or What does fourteen time three hundred sixty nine equal?"

Some examiners may ask an embarrassing personal question, such as, "Did you masturbate this morning?" Others merely threaten by saying, "OK now Karen, I'm going to ask you a very personal and embarrassing question, in your entire life have you ever...?" The polygraph operator never finishes the question; he just records you reacting away in anticipation. Fortunately, only the worst of the verbal voyeurs in the polygraph profession still use this type of control question, so you shouldn't be confronted with it very often.

Here's an update regarding the questions on the LAPD's pre-employment polygraph examination. A recent applicant was asked the following questions (categorized here by type):

Relevant Questions:

Have you stolen more than $400 in cash or property from an employer?

Are you withholding information regarding your illegal drug history?

Are you withholding information regarding a serious undisclosed crime?

Have you physically harmed a significant other during a domestic dispute?

Regarding your background package, do you intend to answer each question truthfully?

Known-Lie Control questions:

Before applying with LAPD, did you ever cheat on a test?

Prior to applying with LAPD, did you ever tell a lie to someone who trusted you?

Prior to applying for this position, did you ever do anything that would cause anyone to question your integrity?

Before applying for this position, did you ever violate any official rules or regulations?

Prior to applying with LAPD, did you ever do anything bad in your life?

Irrelevant Questions:

Are you now sitting down?

Are you now in the State of California?

Before applying with LAPD, did you attend High School?

Is this the month of April?

Is this the year of 2009?

THE (POLYGRAPH) TEST PHASE

The polygrapher will put a blood pressure cuff around your arm, metal contact bars on your ring and index fingers, and pneumograph tubes around your chest and stomach. (See Exhibit C) The polygrapher tell you the test is about to begin, pump up the pressure in the cuff, and then he will usually ask about ten questions – seven or eight relevant and two or three controls. He will tell you to set up straight, keep your eyes open, (some may say to keep your eyes closed), remain still, and answer "yes" or "no" to each question. The polygrapher will ask the questions at intervals of about 15 to 20 seconds – and he will repeat the question series three times. In other words, he will ask you all the questions, stop and release the pressure in the cuff – then he will start again and ask you all the same questions again two or three more times.

You must learn how to manipulate and control every tracing on the chart during the polygraph test – both the breathing or pneumo tracing and the blood pressure or cardio tracing.

I WILL DESCRIBE HOW TO PHYSICALLY MANIPULATE ALL THE TRACINGS ON THE POLYGRAPH CHART, BUT AFTER MUCH TESTING IN MY PRIVATE POLYGRAPH PREPARTION TRAINING, I HAVE CHANGED MY TECHNIQUE, AND NOW RECOMMEND THAT YOU JUST USE THE MENTAL IMAGERY INSTEAD OF PHYSICALLY MANIPULATING THE REACTIONS. YOU WILL SEE ON THE ONLINE VIDEO/DVD THAT THE MENTAL IMAGERY LOOKS MUCH MORE NORMAL AND NATURAL THAN THE PHYSICALLY MANPULATED REACTIONS.

Your breathing, or pneumo tracing, is recorded by the pneumograph tubes which are placed around your chest and stomach. (See exhibit C) When you inhale or breathe in, these tubes expand and the pneumo pens on the chart go up, when you exhale or breathe out, the tubes deflate and the pneumo pens go down.

The polygraph operator is constantly alert of a person who is controlling his breathing. (See Exhibit D) You will notice the difference between the normal and controlled breathing pattern. The controlled breather shows his attempt to control by consciously thinking of his breathing only to the point that he inhales and exhales, he breathes in and immediately breathes out, showing a jagged edged tracing. You should show a normal breathing pattern on all the questions except the control questions. I don't want you to be obvious about this, I simply want you to breathe the way you normally do - whatever is normal for you

After you answer the RELEVANT questions, your breathing should appear even and restful. You have a pattern for a normal breathing if you simply breathe as though you are asleep and you are not aware of your breathing – picture yourself on a beach watching the waves gently rolling into the shore – just relax. This normal breathing pattern is what the polygraph operator would expect to see from a cooperative, truthful person. Remember: (1) your breathing is recorded on the polygraph chart by the pneumo pens; (2) you must avoid a jagged edged breathing pattern, and (3) breathe as though you are breathing in a normal relaxed manner. The best way to do this is when you hear a relevant question, just think about the most relaxing thing you can imagine and picture yourself doing it. I don't want you to try to make your breathing look "normal" - I just want you to let it be normal. Just relax. Do not internalize the questions – don't think about what the polygrapher is asking. You are to listen to the questions only to determine whether they are relevant or control and if they are relevant, your only job is to keep your mind in a calm, relaxed state. This is just a game - relax and play your part.

After you answer a CONTROL question, you must show a breathing or pneumo reaction. Exhibit E shows the five common pneumo reactions. These are easily produced by simply using the mental imagery – thinking of something frightening – don't try to memorize these reactions just know that they will be produced by mental imagery as easily as trying to duplicate them. You don't have to be perfect – just show a slight change in your breathing pattern after you answer the control question – even a slight change looks big on the polygrapher's chart.

PRACTICE until you can do it without being obvious – subtlety is the key! Study the explanation of the reactions shown in Exhibit E, watch the video/DVD, and practice until you can do it without making it obvious that you are manipulating the reaction.

Figure 1 depicts the most common reaction seen in the pneumo tracing. This reaction is manipulated by duplicating the pattern shown. Breathe by the numbers again. (1) Inhale about one-third the normal amount of air, hold slightly, and exhale slowly, showing no jagged edges, (2) inhale again, this time inhaling about two-thirds the normal amount of air, exhale slowly, (3) inhale and exhale the normal amount of air, (4) inhale again, this time inhaling just a little more air than normal, and exhale slowly. You now take two deep breaths, and resume your normal breathing pattern.

The pneumo reaction in figure 2 is manipulated by inhaling more than you exhale each time in a series of five small breaths until, with your last breath; you fill your lungs with slightly more than the normal amount of air, just like you are frightened and gasping for breath. You then take two deep breaths and resume normal breathing.

Figure 3 is too obvious so don't use it. It is true that it is the easiest, but it is also the least desirable.

Figure 4 illustrates still another pneumo reaction which is manipulated by simply inhaling a normal amount of air and then taking a series of five to seven shallow breaths with your lungs partially full.

Figure 5 is a variation of 4 except that you take five to seven shallow breaths with your lungs almost empty.

That is all there is to controlling your pneumo tracing on the polygraph chart. It doesn't have to be perfect; you are basically just showing a departure from your normal breathing pattern after you answer a control question. You are showing the polygraph operator that the control questions bother you and the relevant questions don't.

NOTE: WHILE I HAVE EXPLAINED HOW TO PHYSICALLY MANIPULATE THE BREATHING, IT IS MUCH BETTER TO JUST USE THE MENTAL IMAGERY – THINK OF SOMETHING FRIGHTENING AND YOU WILL PRODUCE THIS VERY SAME BREATHING REACTION WITHOUT ANY PHYSICAL MANIPULATION AND IT WILL LOOK MUCH MORE NATURAL.

Now that you have mastered the manipulation and control of the pneumo tracing, believe it or not, you have mastered the most difficult part of the "Sting Technique". I told you the polygraph exam was a joke!

OK, so you can now manage to show a "normal" breathing pattern and a breathing "reaction", but what about your blood pressure? Please don't take drugs, they will only make you easier prey for a skilled interrogator, and don't put a tack in your shoe, you will only hurt your foot. Just beating the operator is not enough, you have to "sting" him - and remember, a sting is when you con a con man and he never knows he has been conned. If you are going to sting, you must use your stinger. Please turn to Exhibit F as we discuss a well-known phenomenon associated with your stinger or more correctly your anal sphincter muscle. The anal sphincter muscle is the ring-like muscle that surrounds the lower bowel opening. I became well acquainted with the phenomenon known as the "pucker factor" during my military and police careers. The pucker factor is simply a physiological reaction to fear. For example, every time a gun was fired at me, the pucker factor got very high, and anal sphincter started pinching holes in my underwear.

What happens if you are not really frightened, but you "pucker up" just like you do when you really are? Does it look like a blood pressure increase on the polygraph chart, or as polygraph operators say, a cardio reaction? Yes, if you tighten up your anal sphincter muscle, like you are trying to stop a bowel movement, you can cause a magnificent increase in the cardio tracing immediately. The anal sphincter muscle, when tightened or puckered up, causes a rise in the cardio tracing which leads the polygraph operator to believe you have had a really significant cardio reaction. The anal sphincter muscle when tightened or tensed, manipulates a rise in the cardio tracing that duplicates a cardio reaction. A cardio rise, or reaction, can be controlled by relaxing the tension. You simply tighten your anal sphincter muscle and the cardio pen goes up (manipulated reaction), or relax your anal sphincter and the cardio pen goes down (controlled reaction). This muscle is not only capable of manipulating and controlling the cardio tracing, it has the added advantage of being concealed from the polygraph operator.

CAUTION, a few years ago the polygraph industry came up with a "sensor pad" that you sit on while taking the test. I have this "motion sensor pad" attachment on my own computerized polygraph instrument - this is the polygraph instrument I use to train people when they come to me for personal polygraph test preparation training. It really doesn't detect anything, but often people overdo it with the anal sphincter and it looks too obvious. **That is why I now teach people to just use the mental imagery – think of something frightening and it will cause a subtle increase in blood pressure and pulse rate that is just as effective as the tightening of the anal sphincter muscle, AND IT IS MUCH MORE NATURAL LOOKING - see the mental imagery reactions on the online video or DVD.**

The GSR (Galvanic Skin Response), or sweat activity is relatively unimportant and will be both manipulated and controlled to some degree by the manipulation and control of the breathing and blood pressure. If you manipulate and control the pneumo and cardio tracings, the GSR will mirror these responses.

Timing is very important in the manipulation and control of your chart tracings; you must know when to show a manipulated reaction and when not to show a reaction. Your reaction should last about 7 to 9 seconds, and the cardio rise should peak at about 4 to 5 seconds. In other words when you are duplicating your pneumo reaction you should also duplicate a cardio rise by tightening your anal sphincter muscle gradually until you are about halfway through the breathing pattern and then gradually relax it so that the peak blood pressure increase is in the middle of the breathing pattern. You must do this after you have answered the control question. You are to react to the control questions, and your best bet is to mix up the pneumo reactions, using different ones each time.

All I am teaching you to do is to duplicate the physiological response to fear, but you must be able to do it on demand and at the appropriate time. When you are frightened you breathe in shallow, erratic, panting gasps, your anal sphincter muscle puckers up, your blood pressure increases, and you start to sweat. You will also show all these things on the chart when you think of something frightening.

This brings us to our primary method of manipulating reaction – the mental imagery reaction. This is the one I recommend most because it is foolproof and looks like a natural reaction on the polygraph chart. This reaction is manipulated by simply thinking about a math question such as, "What is 278 divided by 13?" Or you can count backwards from 700 by 3s. Most people think about the thing they fear the most or the most frightening thing that has ever happened to them - and they put that image in their mind after they answer the control questions. That frightening mental image causes a great reaction to the control question!

CAUTION!!! It is important that you fine-tune the "sting". The polygraph operator will usually ask the same questions over 2 or 3 times. And I want you to manipulate a reaction to the controls using only the mental reactions.

NOTE: I suggest using the mental imagery to cause a reaction to a control question on all the control questions because it works the best and all the polygrapher will see is a perfect natural reaction. I have given personal instruction to many people, and everyone has been able to manipulate a very good reaction to the controls by simply using the mental imagery. Most people think about the thing they fear the most or the most frightening thing that has ever happened to them - and they put that image in their mind after they answer the control questions. That frightening mental image causes a great reaction to the control question! And, of course, on the relevant questions, you simply concentrate on the most relaxing thing you can think of. As a matter of fact, the mental imagery works best, and it is absolutely natural.

IF YOU ARE NOT SURE IF A QUESTION IS RELEVANT OR CONTROL, DON'T REACT TO IT. IT IS BETTER TO REACT TO ALL THE CONTROLS - ESPECIALLY THE KNOWN-LIE CONTROLS BUT, AS LONG AS YOU REACT TO ONE OR TWO OF THE CONTROLS YOU ARE GOING TO BE IN GOOD SHAPE.

TIMING IS IMPORTANT! You must show both a breathing and blood pressure reaction simultaneously when you answer the CONTROL questions, and you must appear calm, relaxed, and breathing normally when you answer the RELEVANT questions.

When you are employing the Sting Technique, your polygraph examination is in fact your examination. Your interests alone dictate whether the questions are relevant or control. Always bear in mind that the purpose of the test is to elicit information from you. The purpose of the Sting Technique is to allow you to control the amount of information you give, and to teach you to manipulate and control your reactions so the polygraph will verify your truthfulness.

THE VSA, AND CVSA

The VSA, (VOICE STRESS ANALYZER), or the CVSA, (COMPUTER VOICE STRESS ANALYZER), as the name implies simply tries to detect deception by the tremor in your voice. It is even more of a joke than the polygraph. It is easily manipulated and controlled. They use the same type of relevant and control questions as the polygraph. You just answer the relevant questions in a monotone and you give your answer after you have INHALED a normal breath. In other words you answer the relevant question and then exhale. And when you answer the control questions you tighten the anal sphincter muscle and answer the question after you have EXHALED your breath. In other words you answer the control questions after you have exhaled all your breath.

RELEVANT QUESTIONS - answer in MONOTONE AFTER YOU INHALE a normal breath. CONTROL QUESTIONS- tighten up your anal sphincter and answer after you have expelled all the air from your lungs. Don't be obvious about it - and spend some time practicing so you can do it right.

THE "POST-TEST" INTERROGATION

Your mastery of the Sting Technique is almost complete; the only area left unexplored is how to conduct yourself during the interrogation. Remember, the whole test is nothing but an interrogation. The sole purpose of the polygraph test is to get incriminating information from the subject. The polygraph operator is usually an expert interrogator, and, like most interrogators, he relies on his ability to con you or scare you. Do not allow him to do either, concentrate on what you are saying, and what you are doing. Stay alert, remember the polygraph is a joke, and the polygraph operator is playing a con game, a game you will win if you use the Sting Technique correctly. If the polygrapher suspects you of deception (and sometimes even if he doesn't), he will tell you the polygraph indicates deception and try to get some admissions from you. Interrogation techniques vary, but typically, the polygrapher will ask you to explain why you reacted to a certain question. Here are some examples of "tried and true" interrogation techniques.

These interrogation approach phrases are taken from are from the DoDPI "Interview and Interrogation" handbook ·They didn't bring me here to ignore my report. The test confirms that you haven't been completely truthful. Your situation will only get worse if we don't get this cleared up. ·The only thing that will help you now is to be completely truthful.

When a person hides something or lies they usually regret it later on when the truth comes out... like it will in this situation. ·We've all been in situations when we withheld or told a lie about something that didn't seem too bad. But then, we had to tell another lie and another lie and another until the whole story fell apart. ·It is no longer an issue as to whether you did this or not. The only things left to discuss are why and how you got involved in this matter. In fact it is really an insult to my intelligence for you to tell me that you have been completely truthful here today. ·I promised that I would be honest with you here today [!] and you promised me the same thing. You and I both know that you haven't been truthful now. I could respect you more if you just told me that you don't know how to deal with this... that you don't want to confess. ·If you were to show me a picture of someone close to you, I could never persuade you that it was someone else. These charts are like a picture of truth or deception and we can't change them no matter what we say. ·A lie is like a cancer inside of you that eats away at you and never goes away until it is taken out. Then the body can get well. Raymond J. Weir, Jr., former head of the NSA polygraph program and past president of the American Polygraph Association, has described a favorite NSA "post-test" interrogation approach (Weir, 1974): "We have a standard interrogation procedure where the examiner looks at the charts, looks at the subject, shakes his head, and says sadly, "I'd like to believe you, Mr. Jones. You do sound sincere to me. But how can I believe you, when you don't believe yourself? You can lie to me, and I don't know you well enough to tell. But you can't lie to yourself—and that's what I'm getting on these charts." (pp. 154-55) Veteran polygrapher Leonard H. Harrelson, president of the Keeler Polygraph Institute in Chicago since 1955, suggests this outrageous ploy in describing what he terms the "unexpected" or "shock" approach (Harrelson, 1998):..."the imagination and the role-playing ability of the examiner is given free reign. This approach would include such tactics as suddenly shutting off the instrument in the middle of a test, removing the attachments from the subject and requesting that he get down on his knees to join you in praying for his soul and courage to tell the truth. This approach, if used with sincerity and conviction, can carry a tremendous psychological impact on certain subject types".

When a person is lying, they use many different body signals.

A liar is worried about being found out (unless they are psychopaths or good actors), they demonstrated this tension by sweating, sudden movements, minor twitches of muscles (especially around the mouth and eyes), changes in voice tone and speed.

Liars try to avoid detection by over-control. For example, there may be signs of attempted friendly body language, such as forced smiles (mouth smiles but eyes do not), jerky or clumsy movements. The person may also try to hold their body still, to avoid tell-tale signals. For example they may hold their arms in or put their hands in their pockets.

A liar has to think more about what they are doing, so they may drift off or pause as they think about what to say or hesitate during speech.

They may also be distracted by the need to cover up. Thus their natural timing may go astray and they may over- or under-react to events.

Anxiety may be displayed by actions such as fidgeting, looking around the place or paying attention to unusual places.

Deceptive people are worried about being detected - this may be seen in what they do:

Deceptive people:	Do these things:
... are tense because they are worried about being caught or are feeling guilty.	...speak in a high pitched voice ...hesitate. ...stutter. ...have jerky movements.
...can't remember.	...story is inconsistent. ...leave out irrelevant details. ...are vague, or leave out information about times, places and feelings.
...manufacture information.	...often hesitate so they have time to think about what to say. ...or they forget and have to take time to try to remember what they have said.
...don't want to answer the questions.	...cover their mouths. ... press lips together when difficult topics are mentioned.

...try to avoid answering some important questions.	...so they wander away or try to change the subject.
...try to confuse the interrogator.	... they try to give complex answers. ...question the minor details of the question.
...try to appear truthful by acting innocent.	...eyes wide open, with raised eyebrows. ...carefully enunciate speech. ... pout. ...start to cry.
...are worried about what might be asked.	...ramble on to try to use up time. ...try to get emotional to get you off the subject.
...are worried about making admissions that would incriminate them.	...choose their words carefully. ...pause before answering. ...give short answers.

...fear making eye contact because they feel it will give them away.	...avoid eye contact, turn their eyes or head down or away. ...glance away when they start to lie. ...blink and/or rub eyes.
...the fear being detected.	...so they try to say as little as possible. ...try to get away from or change the subject. ...repeat back your words with a denial. ...make exaggerated statements about being truthful.
...use controlled language.	...try to be very precise. ...generalize, using words like "always", "nobody", etc.). ...do not use contractions (saying 'do not', vs. 'don't').

...try to control their body language.	...hold the body still and rigid. ...smile with the mouth but not the eyes. ...try to act innocent by using exaggerated movements.
...when they can't control their body language.	... eye pupil dilation. ...briefly shrug and grimace ...fidgeting movements with hands and feet.
...become nervous and speed up.	...talk faster. ...blink more. ...swallow more. ...move faster.
...think they are threatened.	...attack, defend or deflect. ...place barriers in front of them, from arms to books to tables.
...liars need time to think.	...stall, by repeating the question back to the interrogator ...talk very slowly and deliberately.

...try to remain neutral and tell a story.	...they look up and/or up and right as if they are remembering and describe things as if they are looking at a picture.
...try to distance themselves from what is being asked.	...don't use "I" in their answers. ...say, "believe it or not" or "you probably won't believe this, but...". ...flatly deny any involvement at all.
...Start to sweat.	...their skin gets redder or damper. ... they rub their palms and head, or the neck and nose.

These things are important to keep in mind during your interview/interrogation so you don't give yourself away.

Always appear cooperative, act sincere, use plenty of eye contact, stay alert, concentrate on what you are doing, and never exhibit any hostility, arrogance, or fear, and you can counter any interrogation technique the examiner can throw at you. Just continue to maintain that you have told the complete truth. Volunteer to take the test again if that would help to prove your truthfulness.

Now, if you can add one more finishing touch, it is to look the examiner in the eyes when you talk to him. If you can't do that, focus on the bridge of his nose, right between his eyes, it will have the same effect, which is to prove to him you are truthful. Do not tolerate physical violence, but do not confuse noise with violence. Hang in there unless he gets physical. The polygraph operator who is not well versed in the fine art of interrogation, (or those who have gone crackers from too much hog killin'), relies more on noise and threats than on intelligence, and may try to intimidate you with loud yelling, hostile accusations, and threatening gestures. He is easily manipulated, do not respond to him in the same manner, remain calm, appear cooperative, and act like you are confused by his anger. He will wear out and leave you alone if you refuse to respond. Stick to your story; let him rant and rave, while you think about cartoons or something pleasant. Remember that the entire polygraph examination must be manipulated and controlled if your sting is to be complete. That includes the pre-test interview, the test itself, and any post-test interrogation you may encounter.

PRACTICE EXAMINATIONS

Please practice before you go one-on-one against a professional! To successfully duplicate a "truthful" polygraph chart tracing, you must be prepared to deal with any polygraph technique you may encounter. The following are three practice examinations; each employing one of the three different techniques previously discussed. You will be told how to manipulate and control your reactions as we examine each of these tests. If you are prepared to handle these, you can handle any type of polygraph test, but please practice.

The first test is called a specific issue test. As the name implies it deals with a specific issue, such as theft, murder, rape, etc. You simply change the relevant questions to suit your situation. This one deals with a "leak" of information. Other specific issue tests are called by other names but they will have this same structure they are sometimes called CI-, or counterintelligence-scope polygraph examinations deal which with whether or not you have disclosed classified information to unauthorized persons, had unauthorized contacts with representatives of a foreign government, or been directed by someone to seek employment with a U.S. agency, or "lifestyle" polygraph examinations which deal with more than the subject matter of CI-scope polygraph examinations and will probably include questions about drug use, undetected felony crimes, and, in some cases, sexual conduct.

1. Do you live in the United States?
2. Is your name Harry Jones?
3. ® Do you know for sure who gave that information to the New York Times?
4. ® Did you give that information to the New York Times?
5. ® Besides what you've told me, have you ever stolen anything else?
6. ® Can you name the "informed sources" quoted by the New York Times?
7. ® Have you lied to me about leaking classified information?
8. ® Have you lied to me on any of these questions?
9. ® Besides what you've told me, have you ever lied to anyone else who trusted you?
10. ® Are you now concealing any information whatsoever about the leaks of classified material?

With questions one and two, (IF THE FIRST TWO QUESTIONS ARE THE SAME OR SIMILAR TO THOSE LISTED ABOVE, THEY ARE CALLED INTRODUCTORY QUESTIONS –

YOU DO NOT MANIPULATE A REACTION TO INTRODUCTORY QUESTIONS), the polygraph operator is allowing you time to become accustomed to the sound of his voice, this gives you time to relax and just breathe calmly.

Question three is your first relevant question. You should listen to the questions only to determine whether they are relevant or control, don't let the polygraph operator "psyche you out", remember this is just a game, and all you have to do is breathe normally. Question four is the most relevant question. Simply picture the most relaxing thing you can imagine in your mind and you will automatically produce normal breathing pattern on the polygraph chart.

REMEMBER, KNOWN-LIE CONTROL QUESTIONS ARE THE MOST COMMONLY USED, IF THE POLYGRAPH OPERATOR IS USING KNOWN-LIE CONTROL QUESTIONS YOU ONLY REACT TO THE KNOWN-LIE CONTROLS.

Question five and nine are your prime control questions. When you use the frightening image in your mind, the reaction will look just like Exhibit B.

The scenario for the second polygraph exam is as follows: You have applied for a job, one of the few that still are legally allowed to require polygraphs - most police and federal agencies still require a pre-employment exam.

    1.   Is your first name Janine?
    2.   Do you live in Dallas?

3. ® Did you tell the complete truth on you job application?
4. ® Have you ever been arrested?
5. ® Do you drink?
6. ® Besides what you've told me, have you ever stolen anything else?
7. ® Have you ever used or sold narcotic drugs?
8. ® Are you now concealing any information about your previous work record?
9. ® Are you behind on any of your bills?
10. ® Have you told me the complete truth about how much you owe?
11. ® Besides what you've told me, have you ever told any other lies?
12. ® Have you told me the complete truth about your physical condition?
13. ® Is there anything in your background that would disqualify you from getting this job?
14. ® Besides what you've told me, have you ever lied to get out of trouble?
15. ® Have you lied to me about what you have stolen from previous places of employment?
16. ® Is there anything in your personal life that might interfere with your employment?
17. ® Do you gamble?
18. ® Have you lied to me on any of these questions?
19. ® Are you absolutely trustworthy?

REMEMBER, KNOWN-LIE CONTROL QUESTIONS ARE THE MOST COMMONLY USED, IF THE POLYGRAPH OPERATOR IS USING KNOWN-LIE CONTROL QUESTIONS YOU ONLY REACT TO THE KNOWN-LIE CONTROLS.

NOTICE!!! SOME PRE-EMPLOYMENT TESTS NOW HAVE ONLY RELEVANT QUESTIONS WITH NO CONTROLS AT ALL; (But this is very unusual and most will have some sort of control question to which you can manipulate a reaction.)

1) Do you live in California?
2) Do you intend to lie on any of these questions?
3) Have you lied about your qualifications to do this job?
4) Have you lied about your driving record?
5) Have you lied about your criminal or arrest record?
6) Have you lied about your drug use?
7) Have you ever struck a significant other?
8) Have you ever stolen something from an employer?
9) Have you shoplifted anything since age 18?
10) Have you committed any sexual crimes?
11) Have you placed any false information on your application?
12) Have you omitted any information from your application?
13) Have you committed any serious crime?

14) Have you lied about your involvement in any alcohol related crimes?
15) Have you omitted anything that if discovered would prevent you from being hired by this department?
16) Have you lied on any question I have asked you?

IF YOU ENCOUNTER ONE OF THESE, YOU ARE TO SIMPLY BREATHE IN A CALM EVEN MANNER WHEN YOU ANSWER THESE QUESTIONS. (I have only heard of a few of these, and it is absurd that they do it but if they do, it is very simply, just relax and think of a relaxing image in your mind and you will have no problem).

A good way to practice your Sting Technique is to have a friend read the questions to you, or records them on a tape recorder, allowing an interval of about fifteen to twenty seconds between questions. You should answer the questions aloud with a yes or no, while at the same time manipulating or controlling your reactions to these questions. It is very important to spend as much time as is necessary to feel comfortable before you take the test. The more often you hear the questions, the less likely you are to react to them on the test. So get the tape recorder and get busy! PRACTICE MAKES PERFECT, SO PRACTICE!!!!!!!

The third practice test is a periodic polygraph examination. This Gestapo-type test is usually given every six months to employees "picked at random" from the work force.

1. Is your last name Jones?
2. Were you born in Oklahoma?
3. ® Can you name anyone in the company who is stealing?
4. ® Have you stolen anything from the company in the past six months?
5. ® Besides what you've told me, have you ever lied to a supervisor?
6. ® Have you violated any of the company's rules and regulations?
7. ® Are you working with anyone to steal from this company?
8. ® Have you devised a plan to steal from this company?
9. ® Besides what you've told me, did you ever lie to cover something up?
10. ® Do you use or sell drugs?
11. ® Have you ever used drugs or alcohol on company time?
12. ® Are you covering up for anyone who is stealing from this company?
13. ® Besides what you've told me, did you ever reveal anything told to you in confidence?
14. ® Have you lied to me on any of these questions?

You should have no difficulty recognizing these control questions! REMEMBER : IF THE FIRST TWO QUESTIONS

ARE THE SAME OR SIMILAR TO THOSE LISTED ABOVE, THEY ARE CALLED INTRODUCTORY QUESTIONS – YOU DO NOT MANIPULATE A REACTION TO INTRODUCTORY QUESTIONS.

Remember to manipulate a reaction to the control questions, and show no reaction whatsoever to the relevant questions. The examiner will ask probably

ask all of the questions over again 3 or 4 times, each time the questions may be in a different order.

REMEMBER, KNOWN-LIE CONTROL QUESTIONS ARE THE MOST COMMONLY USED, IF THE POLYGRAPH OPERATOR IS USING KNOWN-LIE CONTROL QUESTIONS YOU ONLY REACT TO THE KNOWN-LIE CONTROLS.

NOTE: I suggest using the mental imagery to cause a reaction to a control question on all the control questions because it works the best and all the polygrapher will see is a perfect natural reaction. I have given personal instruction to many people, and everyone has been able to manipulate a very good reaction to the controls by simply using the mental imagery. Most people think about the thing they fear the most or the most frightening thing that has ever happened to them - and they put that image in their mind after they answer the control questions. That frightening mental image causes a great reaction to the control question! And, of course, on the relevant questions, you simply concentrate on the most relaxing thing you can think of. As a matter of fact, the mental imagery works best, and it is absolutely natural.

Often, the polygraph operator will tell you not to answer the questions aloud but to remain silent, (some will ask you to nod or shake your head yes or no). You are to manipulate a reaction to the control questions and remain calm on the relevant questions just like you would if you were answering aloud.

Also he may ask you to deliberately lie on all the relevant questions - DO NOT manipulate a reaction to ANY of the questions when he tells you to do this - DO NOT react to the relevant questions or the control questions.

If you are not sure if a question is relevant or control - don't react to it! There will be at least one or two that you will recognize and that will be sufficient to pass the test. One more quick tip, the polygraph operator may ask you which question you remember out of them all, always say you remember the control question, because that indicates to him it troubles you the most. Never indicate by words or actions that the relevant questions caused you any trouble at all, and never ask him how you did on the test, just assume you have passed, thank him for his time and leave the room.

I have one more bit of instruction for you to remember. Often, especially in pre-employment tests, the polygraph operator will tell you that you had a problem with one or two questions. Usually your "problem" was with the drug questions. He will tell you to tell him what you are withholding or to go home and write it all down. He will schedule you for another test later that will concentrate on this issue. Don't panic!!! The sky is NOT falling. This is just another interrogation trick. Stick to your story and go back and do the "sting" on the next test.

The polygraph profession has accused me of having a myriad of sordid motives for writing this little manual, perhaps their anger stems from the fact that the only power the have is derived from the fear and ignorance of their victims. I only hope I have been successful in tipping the balance of power from the terrorists to the victims by telling some of the tricks of this terrible trade. I have done a great deal to outlaw the use of this insidious Orwellian instrument of torture in the private sector, but much more needs to be done. I look forward to the day when the polygraph test will just be a bad memory. As to my ex-colleagues' criticism, I would offer this thought for consideration. It would, after all, be the ultimate irony for a "lie detector" operator to object to the truth.

By the way, I have made polygraph operators so paranoid that they are now asking everyone they test whether they have looked at my website and read my manual, (some will even hold up their personal copy and show it to you to try to convince you they know all about me). The question about whether you have prepared for the test or are trying to beat it is often asked on the polygraph test itself! The mere fact that they accuse everyone proves they don't see anything but a truthful chart tracing - they are just fishing, trying to bluff you into making an admission. Don't admit anything, just stick to your story, keep telling the polygraph operator that you have told the truth, and tell him that if he needs to run another test to see that to go right ahead!

What a joke these guys are - over 100,000 people have used my technique to pass their polygraph tests and there are literally hundreds who use the "sting" technique daily and they haven't caught even one of them. The only ones they "catch" are those that use some of the other crap put out by guys whose only experience with the polygraph is that they once flunked a test.

One more reminder, PLEASE PRACTICE!!! That is the only way to be properly prepared and also the only way to overcome nervousness. Get your tape recorder, record the questions and practice answering aloud yes or no. Practice until you are able to show your normal breathing patterns on the relevant questions and your manipulated reaction on the control questions. The more you hear the questions, the less likely you are to have a reaction on the relevant questions and the more you practice your mental imagery the easier it gets. And before you answer the questions, label them in your mind as either relevant or control and then just think of the most relaxing thing you can imagine on the relevant and the most frightening thing you can imagine on the controls – that is the perfect way to produce a perfect chart. So PRACTICE, PRACTICE, PRACTICE!!!

If you have any questions at all about the "Sting" technique, call or e-mail me. I WILL get you ready and I WILL be your very own personal DRILL INSTRUCTOR. I want you to be prepared. I want you to succeed. I do require that you read the manual over again at least 4 times so that you have a thorough understanding of the information before you call or email me with any questions. You are wise to prepare yourself because just telling the truth is no guarantee of passing the test.

One last word of advice: It is up to you to make sure you are PROPERLY PREPARED! What does it mean to be PROPERLY PREPARED? It means you are able to do exactly what the manual and dvd tell you to do, and do it the right way! You must study the manual, you must practice, and you must do everything the manual and dvd tell you to do – nothing more and nothing less! As a matter of fact, it is entirely your responsibility and yours alone to make sure you are PROPERLY PREPARED. I can give you the information, but it is up to YOU to do it PROPERLY!!!

Review:

Remember, there are two types of questions on the polygraph test – RELEVANT & CONTROL

You are to listen to the questions only to determine what they are and what you are to do in response to them.

RELEVANT QUESTIONS – Answer them and just RELAX – think of the most relaxing place you have ever been and go there in your mind after you answer the relevant questions.

CONTROL QUESTIONS – Answer them and manipulate a reaction after your answer. And just do the mental imagery after you answer them. Think of the most frightening thing that has ever happened to you and relive that experience in your mind after you answer the control questions.

Review the KNOWN-LIE CONTROL QUESTIONS – and memorize them so you can recognize them immediately – they are the most commonly used control questions.

LABEL THE QUESTIONS IN YOUR MIND BEFORE YOU ANSWER THEM! LABEL THEM AS RELEVANT OR CONTROL AND DO THE APPROPRIAT E MENTAL IMAGERY.

It really is that simple! And you will pass if you just follow these simple instructions!

A NOTE TO THE READER

Normally this point Doug Williams would offer his services in a one on one environment to assist those unfortunate people that are facing a pending polygraph siege. Regrettably, at this time Doug is unable to do so. The reason for this is, after the government incarcerated Doug, they decided that as part of his release he should be placed on probation until July 27, 2020. As part of that probation he is restricted from his life's work of assisting people through personal instruction in order to keep them from being victimized by the polygraph industry.

However, after July 27, 2020 Doug will be released from the bondage of probation and will resume his very much needed teaching.

In closing, here is a question. If the polygraph actually works, why would the government, or polygraph operators in general, want to keep Doug Williams restricted at all? Simple. Because the polygraph is not a lie detector as Doug Williams has proven so many times.... and will do so again

COPYRIGHT © 1979 - 2012 Doug Williams,
Sting Publications All Rights Reserved.

EXHIBIT A

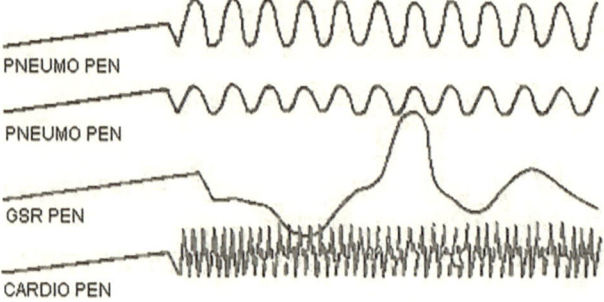

Pneumo Pens - record your breathing or respiration. When you inhale, the pens go up. When you exhale, the pens go down.

GSR Pen - records increase or decrease in your sweat activity or perspiration.

Cardio Pen - traces heartbeat and records changes in your blook pressure and pulse rate.

EXHIBIT B
A Classic Lying Reaction

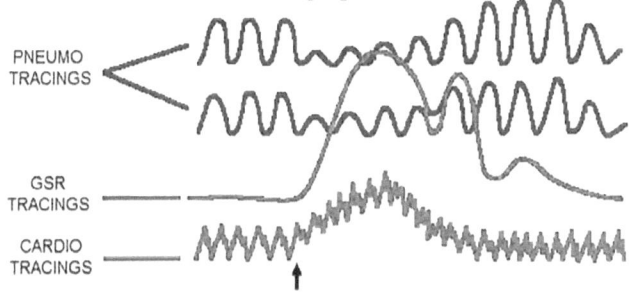

PNEUMO TRACINGS

GSR TRACINGS

CARDIO TRACINGS

THE POINT WHICH THE SUBJECT ANSWERED A QUESTION.

EXHIBIT C

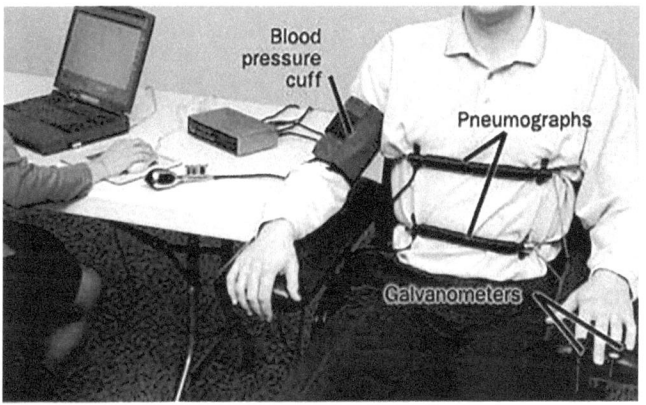

EXHIBIT D

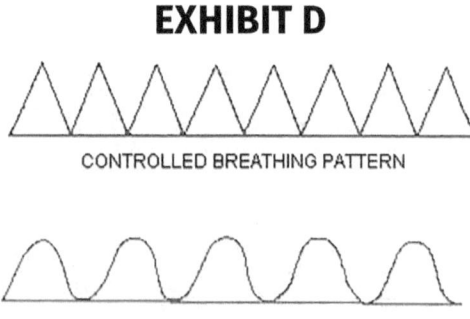

CONTROLLED BREATHING PATTERN

NORMAL BREATHING PATTERN

EXHIBIT E
PNEUMO REACTIONS

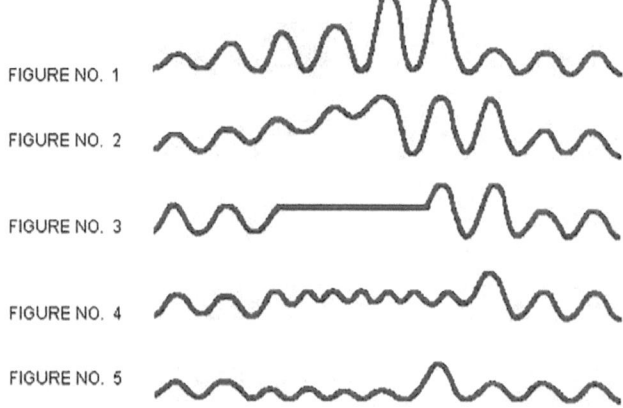

FIGURE NO. 1

FIGURE NO. 2

FIGURE NO. 3

FIGURE NO. 4

FIGURE NO. 5

EXHIBIT E
CARDIO REACTIONS

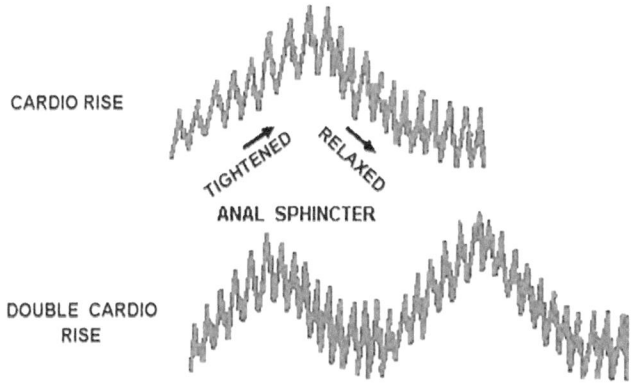

* PRE-EMPLOYMENT PRE-TEST QUESTIONS

1. Is the above name your true legal name?
2. Have you ever used any other name?
3. How many times have you been married?
4. What us your true date of birth?
5. Do you have a legal right to work in this country?
6. Are you skilled or trained in any field in which you could make more money than this job pays?
7. If offered this position, will you accept it?
8. If you accept this position, will you stay with this agency for at least 2 years?
9. Have you had any conflicts with your family because you want this job?
10. Have you ever before been asked to take a polygraph examination?
11. Have you ever failed a polygraph examination?
12. Have you placed any false information on your employment application or personal history background forms?

13. Have you omitted any information requested on your employment application or personal background forms?
14. When you left high school did you receive a graduation diploma?
15. Have you completed a law enforcement academy?
16. Have you ever failed or dropped out of a law enforcement academy?
17. Would you have any reason to be concerned about an investigation into your past work record?
18. Were you ever fired from a job?
19. Were you ever asked to resign from a job?
20. Did you ever leave a job to avoid being fired?
21. Have you ever left a job without giving proper notice?
22. Have you ever been accused of misconduct at a place of employment?
23. Have you shown the true and complete reasons for leaving each of your previous jobs?
24. Did you ever leave any job with hard feelings toward the management or coworkers?
25. Do you think you could return to work for all your former employers?
26. In the past year, how many times have you been late to work?
27. Have you ever received a written or verbal reprimand?
28. In the past year, how many unauthorized days of work have you missed?

29. Have you ever worked at any law enforcement agency in any capacity?
30. Have you ever before applied at any law enforcement agency for any type of job?
31. Have you ever been rejected by any law enforcement agency for any type of job?
32. Were you turned down as unacceptable by the military or draft board?
33. Are you currently registered for the draft?
34. Have you ever served in any branch of the Armed Forces?
35. Would you have any reason to be concerned about an investigation into your arrest record?
36. As a juvenile or adult, have you ever been arrested?
37. Have you ever been held, detained, questioned, or taken into custody for any reason?
38. Have you ever had a warrant issued for your arrest?
39. Are you now wanted for any reason by any law enforcement agency?
40. Have you ever been a suspect in any criminal investigation?
41. Have you ever been charged with a crime?
42. Have you ever been present when anyone else committed a crime?
43. Other than minor traffic matters, have you ever been fined by a court?
44. Have you spent any time, either as a juvenile or an adult, locked up in a jail?

45. Have you ever falsified an income tax form?
46. Have you ever falsified an insurance claim?
47. Have you ever collected unemployment or welfare benefits when you were not entitled to it
48. Have you ever stolen a motor vehicle?
49. Have you ever shoplifted anything from a store?
50. Have you ever been sent to jail over anything involving a motor vehicle?
51. Have you ever deliberately damaged or destroyed any property or committed an act of malicious mischief?
52. Other than from an employer have you ever stolen anything?
53. Have you committed any serious undetected crime?
54. Have you ever made serious plans to commit: A. Burglary, B. Rape, C. Robbery, D. Murder, E. Arson, F. Theft
55. Have you within the past 5 years done anything at all that you could have been arrested for doing?
56. Would you have any reason to be concerned about an investigation into your moral background?
57. Have you committed any type of sexual crime?
58. Since you were 18 years old, have you thought about committing some type of sexual crime?
59. Have you ever paid for sex?
60. Have you ever received payment for sex?
61. Have you ever sexually molested a child?
62. Have you ever committed a sexual act in public or an act of indecent exposure?

63. Would you have any reason to be concerned about an investigation of any illegal drug use by you?
64. Have you ever smoked marijuana in your life?
65. Have you ever used: A. cocaine, B. LSD, C. PCP, D. Magic mushrooms, E. Hash, F. Meth, G. Uppers, H. Downers, I. Any illegal drugs?
66. Have you ever worked under the influence of illegal drugs?
67. Have you ever ingested a substance you thought was an illegal drug and then found out it wasn't?68. Have you ever illegally misused or abused any prescription?
69. Of your own knowledge, do your present circle of friends and acquaintances use any type of illegal narcotics, pills or drugs?
70. Within the past 3 years, have you been in the presence of anyone else using illegal drugs?
71. Have you ever illegally purchased any type of narcotic, pill, or drug?
72. Have you ever sold any type of illegal narcotics, pill, or drug?
73. Have you ever cultivated marijuana?
74. Have you ever been involved in the manufacture of any drug?
75. Have you ever been the "middle man" for a drug deal?
76. Has anyone other than a medical person injected anything into your body?
77. Do you object to others using narcotics?

78. If employed as a peace officer, would you arrest a friend, if you came upon that friend using narcotics or any drugs?
79. Would you have any reason to be concerned about an investigation concerning your honesty?
80. Have you ever stolen any money from a place where you worked?
81. Have you ever borrowed money from an employer and not paid it back?
82. Have you ever embezzled any money from an employer?
83. Have you ever stolen any merchandise or property by false representation?
84. Have you ever taken any property that didn't belong to you from a place where you worked?85. What is your total indebtedness?
86. Could you successfully manage your financial affairs on the salary this job offers?
87. Have you ever had a debt turned over to a collection agency?
88. Have you ever been late in paying rent?
89. Has your salary ever been garnished?
90. Have you ever had purchased goods repossessed?
91. Have you ever filed bankruptcy?
92. Have you ever avoided paying any lawful debt by moving away?
93. Have you ever been late paying your taxes?
94. Have you ever failed to support any child of yours?
95. Have you ever been late in making child support payments?

96. Have you ever been late in repaying a student loan?
97. Have you ever had a check "bounce"?
98. Have you ever borrowed money to gamble with?
99. Have you ever borrowed money to pay a gambling debt?
100. Do you feel you now have a problem with gambling?
101. Have you ever been the plaintiff or defendant in any civil court action?
102. Do you presently have any civil actions pending in court?
103. Have you had any judgments filed against you?
104. Have you ever been arrested or convicted for any alcohol related crimes?
105. Have you ever worked under the influence of alcohol in violation of company policy?
106. Would you have any reason to be concerned about an investigation into your driving habits?
107. How many traffic citations have you received in your life?
108. Have you ever had a ticket go to warrant?
109. Have you ever had a traffic citation that did not show on your DMV printout?
110. Have you ever been the driver in any traffic accident?
111. Has your driver's license ever been suspended or revoked?
112. Has your auto insurance been placed in the assigned risk pool?

113. Has your auto insurance ever been canceled for cause?
114. Do you now have auto insurance as required by the State?
115. Since being licensed to drive, has there ever been a time when you did not have insurance as required by law?
116. Have you ever caused anyone serious injury by your operation of a motor vehicle?
117. Have you ever caused the death of anyone by your operation of a motor vehicle?
118. Have you ever fled the scene of a hit and run accident?
119. Have you ever driven a motor vehicle while under the influence of: A) Some type of drug? B) Alcohol? C) Combination of above?
120. In the past month have you driven a motor vehicle while under the influence of A) Some type of drug? B) Alcohol C) Combination of above?
121. Have you ever been arrested for driving while under the influence of alcohol or drugs?
122. Would you have any reason to be concerned about an investigation into your personal background?
123. In the past year, have you been in any fight? If so, did you start it?
124. Have you ever struck or injured any person?
125. Have you ever struck someone you were living with?
126. Other than in warfare, have you ever caused serious injury to a human being?

127. Other than in warfare, have you ever been involved in a shooting, knifing or fight, where someone was killed or seriously injured?
128. Other than in warfare, have you ever used any weapon against someone?
129. Other than in warfare, have you ever caused the death of a human being?
130. If it became necessary in the course of your duties to take a human life, would you have any reluctance to do so because of religious or other beliefs?
131. Do you frequently lose your temper?
132. Are you afraid of physical combat?
133. Have you ever fired a firearm?
134. Are you afraid of firearms?
135. Have you ever applied for a permit to carry a concealed weapon?
136. If employed here, would you fear physical resistance by someone you might arrest?
137. Do you feel you can take orders from your superior officers without resentment?
138. Do you have any prejudices relating to race, religion, gender, national origin, or ethnic background?
139. Do you feel your prejudices might affect your ability to perform this job?
140. Have you ever maliciously burned any property?
141. Have you ever turned in a false fire alarm?
142. Have you ever made an obscene or threatening phone call?

143. Have you ever in your entire lifetime done anything at all that you are ashamed of?
144. Is there some undisclosed reason why you want to be a peace officer?
145. Do you know of any reason why you should not be hired by this department for the position you have applied?
146. Is there anything at all in your background that you have not been asked about that might eliminate you from consideration for this job if it were found out?
147. Can you say in complete honesty that you have answered each of these questions truthfully?

Legal Notice: Disclaimer –

The information in this manual is provided for informational purposes only, and is not intended to serve as a source of legal or professional advice. This manual contains the opinions of the author and it in no way guarantees that you will pass a polygraph examination – it is simply the author's opinion that what you have been taught to do here represents the author's best efforts at preparing you for your test. The author and publisher of this manual make no representation or warranties, (expressed or implied) with respect to its accuracy, applicability, fitness, merchantability or completeness. The author and publisher shall in no event be held liable for loss or other damages, including, but not limited to special, incidental or consequential damages.

COPYRIGHT © 1979 - 2012 Doug Williams,
Sting Publications All Rights Reserved.

www.ingramcontent.com/pod-product-compliance
Lightning Source LLC
Chambersburg PA
CBHW021130080526
44587CB00012B/1221